BUILDING ROADS
by
HAND

An introduction to labour-based road construction

Prepared for the International Labour Office
by

J. ANTONIOU • P. GUTHRIE • J. DE VEEN

Longman Group (UK) Ltd
Longman House
Burnt Mill
Harlow
Essex
England   CM20 2JE

© International Labour Organisation 1990

All rights reserved. No part of this publication may be reproduced, stored in a retrieval system, or transmitted in any form or by any means, electronic, mechanical, photocopying, recording or otherwise, without prior written permission of the Publisher.

First published 1990
ISBN 0 582 06060 5

The designations employed in ILO publications which are in conformity with United Nations practice, and the presentation of material therein do not imply the expression of any opinion whatsoever on the part of the International Labour Office concerning the legal status of any country, area or territory of its authorities or concerning the delimitation of its frontiers. The responsibility for opinions expressed in studies and other contributions rests solely with their authors, and publication does not constitute an endorsement by the International Labour Office of the opinions expressed in them. Reference to names of firms and commercial products and processes does not imply their endorsement by the International Labour Office, and any failure to mention a particular firm, commercial product or process is not a sign of disapproval.

Produced by Longman Group (FE) Ltd
Printed in Singapore

Cover Design by Robert Savva

Foreword

• This book serves as a brief introduction to labour-based road construction. It is aimed at supervisory staff and presents the broad technical, organisational and managerial principles of labour-based road construction works. It is not, however, a manual giving detailed information on how to organise and execute such works. It will have to be supplemented by a comprehensive training effort which should be practically oriented. The references listed at the back of the book should be consulted for detailed guidance and information.

Whilst the primary aim of this book is to assist supervisors, it may also be of value to civil engineers who are embarking on labour-based works for the first time, and to students who need a general introduction to the subject.

Table of Contents

1 • About this book

BASIC SKILLS

2 • Length Measurement
3 • Area and Volume
4 • Plan and Cross Section

SURVEYING & SETTING OUT

5 • Methods
6 • Horizontal Alignment
7 • Vertical Alignment
8 • Cross Section

CLEARING

9 • Bush Clearing
10 • Tree Removal
11 • Boulder Removal
12 • Topsoil Removal

EARTHWORKS

13 • Formation
14 • Slotting
15 • Rock
16 • Earth Moving I
17 • Earth Moving II
18 • Compaction

DRAINAGE

19 • General
20 • Road Surface Drainage
21 • Side Drainage
22 • Erosion Control
23 • Turnouts

Table of Contents

24 • Cross Drainage
25 • Fords (Drifts)
26 • Culverts
27 • Vented Fords
28 • Bridges

GRAVELLING

29 • General
30 • Quarries
31 • Haul, Unload, Spread ,Compact

PLANNING

32 • Principles
33 • Daily Work Plan I
34 • Daily Work Plan II

REPORTING & CONTROL

35 • General
36 • Production Records I
37 • Production Records II
38 • Examples of Production Records

STORES

39 • Tools
40 • Materials

STARTING & ORGANISING

41 • Setting up Site
42 • Camp Layout

REFERENCES

The Authors

• **Jim Antoniou** .... is an artist, physical planner and architect. He has extensive experience of developing countries and has led several missions for international agencies in the fields of education, training and infrastructure.

• **Peter Guthrie** .... is a civil engineer and currently an Associate with consulting engineers Scott Wilson Kirkpatrick. This book is based largely on a manual he produced for the Labour Construction Unit in Lesotho following two years work on the project. He has subsequently travelled widely for the ILO in support of labour-based programmes.

• **Jan de Veen** .... is a civil engineer in the Infrastructure and Rural Works Branch of the ILO. He is responsible for roads sector project and training activities, and has had extensive experience with the ILO labour-based construction programmes throughout the world.

Acknowledgements

The production of this publication was made possible with financial support of the Norwegian Agency for International Development (NORAD). Particular thanks are due to Platon Antoniou, Jake Chessum and Jim Antoniou for the illustrations and design of the book with additional drawings by Jim Brown; to John Marshall of the ILO for his advice and assistance on the section on setting-out; and to Hamish Goldie-Scot for his assistance on the editing and to Robert Savva for his work on the production of this book.

About this book

- If you are a supervisor working on a labour-based road construction programme then this book is for you. It is written for the people who have to put into practice the building of roads by labour-based methods. It has been put together by people who have direct experience of supervising this kind of construction.

The book is laid out so that it can be used for easy reference - the main subjects are divided into chapters and each double page spread covers a separate topic.

Above all the book is intended to be a practical guide to people who need to understand the process of road building by labour-based methods .... because they're doing it.

Double page spread number   *Topic title*   *Chapter title*

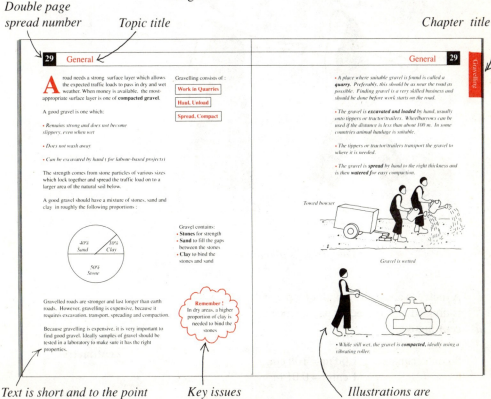

Text is short and to the point for easy reference   *Key issues are highlighted*   *Illustrations are designed to be informative*

2 Length Measurement

Road construction always involves measuring length. Here are some examples:

- How **long** is the road?
- How **wide** is the road?
- How **deep** is the ditch?

Basic skills are:

Length Measurement

Area & Volume

Plan & Cross Section

As you can see, the term **measuring length** is used also for width and depth, as well as **length**.

In this book, length measurements are in metres (m) and centimetres (cm).

One stride is about a metre in length

A **metre** is about the length of a person's stride. A metre is divided into 100 equal parts which are called **centimetres**.

So, in decimals   1.00 m = 100 cm
                     1 cm   = 0.01 m

This is the length of a **centimetre**

1cm

Length Measurement 2

How to measure length

Length is measured accurately with a tape made of either steel or plastic, with length marked on it in metres and centimetres.

Tapes come in different lengths.

The most common are:
3 metres
10 metres
30 metres

Try measuring the length of a pick handle in the following way:

- *First, make sure the tape begins at **zero**, otherwise all your measurements will be wrong*

- *Hold the end of the tape at one end of the pick handle*

- *Then, read the length marked on the other end.*

3 Area and Volume

Area is **length x width**. It is **two** measurements of length multiplied together. This is true for squares and rectangles. A triangle is **half** a rectangle.

Area is measured in square metres, written m^2, or in square centimetres, written cm^2.

Find out the area of this handbook.

- *The **length** is 21 cm*

- *The **width** is 15 cm*

- *Now multiply the length by the width*

- *Area = 15 x 21*
  = 315 cm^2

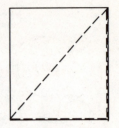

A triangle is half the area of a rectangle ($\frac{1}{2}$ x height x width)

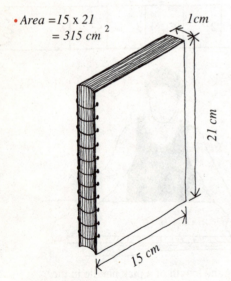

Remember !
Area is length x width

Volume is length x width x height, or area x height. Because volume is made up of **three** measurements of length, it is measured in **cubic metres**, written m^3, or in cubic centimetres, written cm^3.

Remember !
Volume is length x width x height

Look at the drawing of this book again and work out the volume. You already know the area is 315 cm^2, so the volume is 315 x 1 = 315 cm^3.

Area and Volume   3

In the same way try working out the volume of a simple box.

The volume = length x width x height

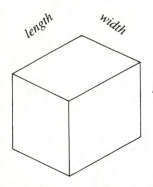

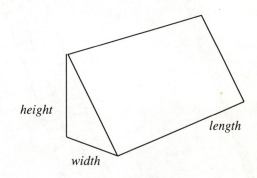

What about the volume of a wedge shape?
The volume is the area of the triangle shape at the end, multiplied by the length.

Volume   = area x length
              = $\frac{1}{2}$ x height x width x length

This information will be very useful to you, for example when working out the amount of topsoil to be removed or excavation to be done.

4 Plan and Cross Section

A plan is what you see if you are looking down from the sky. This view is similar to what a bird would see flying over a road.

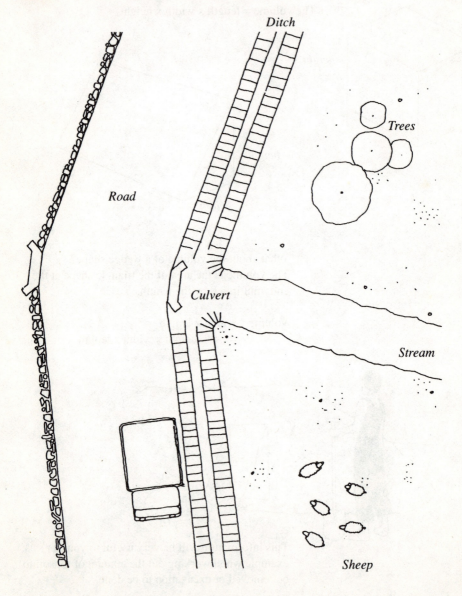

PLAN

Plan and Cross Section — 4

A cross section is a way of drawing something as if it has been cut across. Imagine what a road looks like if it is sliced across the whole width.

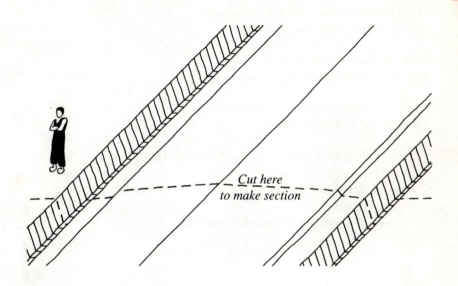

Cut here to make section

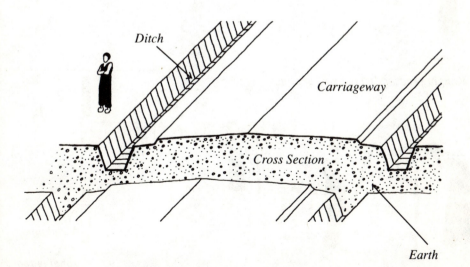

Ditch · Carriageway · Cross Section · Earth

Basic Skills

5 Methods

How to set out the road

• First, decide roughly where the road is to go.
This is called the **selection of the route.**

• Then, set out the **centreline** of the road.
This is called **horizontal alignment.**

• Then, set out the level of the road on the centreline.
This is called **vertical alignment.**

• Then, set out the width of the road and position of the drains from the centreline and set out levels.
This is called the **cross section.**

Tools and equipment for setting out can include :

Surveying and setting out consists of:

| Methods |
| Horizontal Alignment |
| Vertical Alignment |
| Cross Section |

• **Ranging rods** *to set out straight lines*

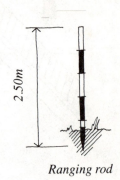

Ranging rod

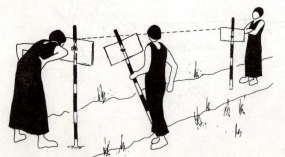

• **Profile boards or boning rods** *to find the correct level*

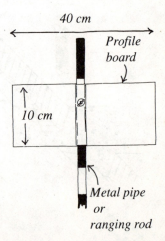

Methods 5

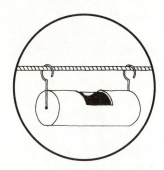

Use nylon string if you can

- **A line level** to find the correct level

- **A tape measure:** you can easily set out a right angle from the centreline through **A** and **B** like this:

From **A** measure 4 metres and where it crosses the centreline mark point **B**.

From **A** also measure 3 metres and mark out an arc on the ground.

From **B** measure 5 metres and where this intersects the arc 3 metres from **A**, put in a peg (at **C**).

The line **A C** is at right angles to the centreline through **A B**.

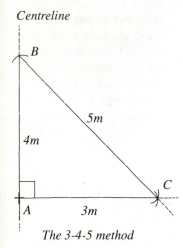

The 3-4-5 method

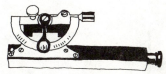

- **An abney level** to measure vertical angles and setting out levels.

- You also need **string** and **wooden pegs**, about 40cm long, to mark out alignment and road levels. Paint the top parts a bright colour so that they can be seen from a distance.

6   Horizontal Alignment

When considering the construction of a road, several horizontal alignments may be possible. A straight and direct alignment from one place to another is not always the best solution.

When selecting the horizontal alignment, try to avoid:

• *knocking down buildings*

• *taking too much valuable farmland*

• *difficult terrain with steep sections, large trees, boulders or swamps.*

It is often possible to choose a slightly longer alignment, which avoids these difficulties, and to build a less expensive and better road

The centreline of the road is set out as a series of **straight** lines connected by **curves**.

Straight lengths are set out by marking the points along the straight lines with ranging rods. These are called the **Intersection Points**. Along a straight line, now mark the centreline at 20 m intervals by guiding a person with a ranging rod so that the rod is in line with the two ranging rods at the intersection points.

> **Remember !**
> The existing track will be well compacted and usually strong. Use this whenever possible except where it is below surrounding ground level or otherwise unsuitable

Intersection point

Intersection point

Horizontal Alignment   6

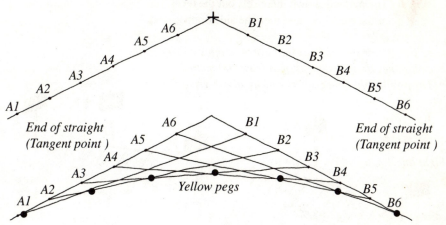

Curves are set out to join two straight sections. First place a peg at the intersection point, which is where the centrelines of the straight sections cross. Now decide where to start the curve at one straight line and place a peg there (A1). Then measure the same distance from the intersection on the other straight line and place another peg there (B6). These two points are called **tangent points**. Now divide the two lines into six equal parts; place a peg at each point and mark them (as shown). Using strings, join the pegs (A6 to B6 and A5 to B5 and so on) as shown in the diagram. Put a yellow peg where the strings cross. Take out the string A6-B6 and put it from A4 to B4; put in another yellow peg where the strings cross. Continue like this until A1-B1. You have now set out the curve with yellow pegs.

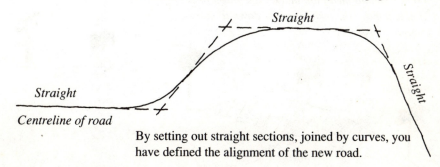

By setting out straight sections, joined by curves, you have defined the alignment of the new road.

7 Vertical Alignment

Vertical alignment sets the level of the road. The method shown here sets out the **level** of the road first, and then the side drains.

• First, fix the profile boards at every ranging rod along the centreline, along distances of at least 100 metres. The profiles should be set at 1 metre above ground level.

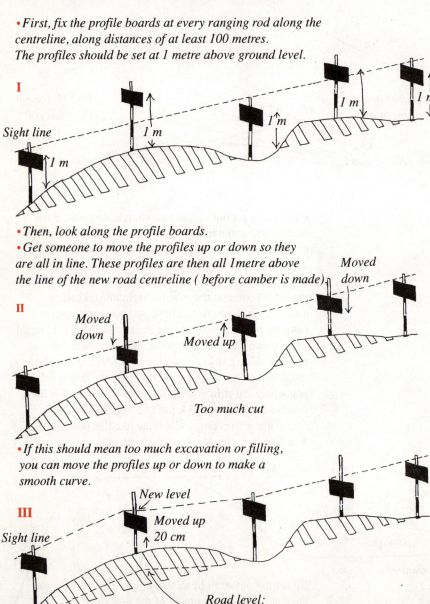

I
Sight line — 1 m, 1 m, 1 m, 1 m, 1 m

• Then, look along the profile boards.
• Get someone to move the profiles up or down so they are all in line. These profiles are then all 1 metre above the line of the new road centreline (before camber is made).

II
Moved down
Moved up
Moved down
Too much cut

• If this should mean too much excavation or filling, you can move the profiles up or down to make a smooth curve.

III
Sight line
New level
Moved up 20 cm
Road level: reduced cut

Vertical Alignment   7

> **Remember !**
> Make sure that the first profile boards are placed correctly. All other levels you set out follow from this

Alternatively, to set out over the crest of a hill, set out a new level by fixing a profile board (C) less than 100 m from A.

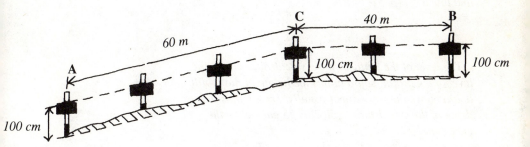

So **sometimes** you need to set the sight line at intervals of **less** than 100 m.

This is how **vertical curves** are formed, and how sharp crests or steep dips are avoided.

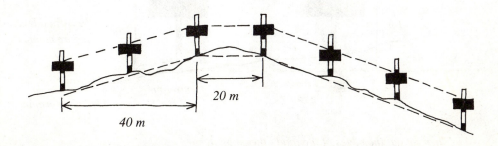

8 Cross Section

Now you have set out the centreline and the levels of the road. Your next task is to set out the **cross section**.

- First, place the profile boards in the middle of the side drains. These should be the correct distance from the centreline and at right angles to it. Use the 3-4-5 method (p. 5) to set out the right angles.

- Then transfer the level from the centreline profile board to the two side drain profile boards, using string and line level (p. 5)

- The level of the bottom of the side drains will be the level of the profile board minus the height of the profile board above the centreline (usually 1m) minus the depth of the side drains, typically 25 cm below the centreline.

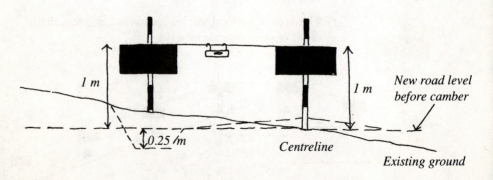

- Then set out the edges of the road , the shoulders.

Cross Section 8

- Then excavate down to the level of the road as set out across the cross section. This means digging or filling until the profile boards are 1 metre above the levelled ground. At the low side of the road you fill only up to the level of the shoulder.

- Finally, excavate the side drains as necessary. In our drawing the bottom of the side drain is 25 cm below the levelled ground. The material from the side drains is used to form the camber of the road.

Remember !
In very flat terrain where drainage is difficult it is better to raise the level of the road centreline by 10 to 20 cm in order to keep the road drained.

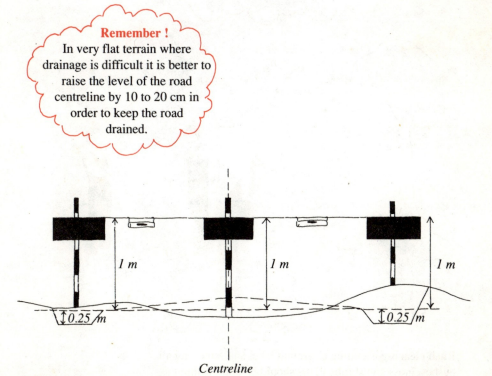

9   Bush Clearing

Clearing is the removal of bushes, trees, boulders and topsoil from the road alignment.

Site clearing is usually the first operation to be done, once the road alignment has been set out. Its purpose is to make construction possible by clearing a path for the road.

Clearing consists of:

Bush Clearing

Tree Removal

Boulder Removal

Topsoil Removal

Remember !
Removing trees and boulders is time consuming. You may also have to pay compensation because every tree belongs to someone. So before starting major clearing work, have you considered the option of changing the road alignment .

Bush clearing is cutting to ground level and removing all bushes, grass and shrubs. This should be done within the road width itself plus usually 5 m on each side. Don't forget to allow for side drains and working space.

Bush Clearing 9

The tools for the job

- Cut dense bushes with a brush hook
- Use a manual plant puller to remove small plants

Brush hook for cutting dense brush

A long stalk with a twig used as a hook

How to do it

- Set out the area to be cleared

- Check on the types of bushes to be cleared

- Decide on the task rate

- Instruct workers which area to clear

- Ensure that workers have enough working space

- Provide the tools

- Now start the work

- Throw bushes outside the cleared area, or heap them in the centre for burning.

Man using plant puller

10   Tree Removal

Before felling a tree, make sure it is absolutely necessary to cut it down. It is true that trees obstruct construction activities. They also prevent a road from drying out quickly after rains and so affect road strength (a dry road is stronger than a wet one).

On the other hand, trees help prevent soil erosion, so they are very useful.

If you leave a tree, do not damage it, or its roots.

Once you decide to cut down a tree, use only experienced workers for the job and keep everyone else well away.

Felling Large Trees
Large trees are felled using their own height and weight.

The safe distance for people is well outside the length of the felled tree

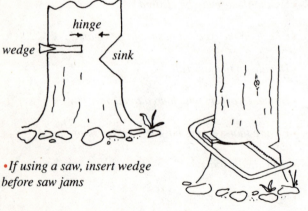

• *If using a saw, insert wedge before saw jams*

• *Cut a felling sink in the direction you want the tree to fall*

• *Make a felling cut with a cross cut saw opposite and above the sink*

• *Make sure the tree falls in the direction you want it to, by using a wedge*

• *Continue saw cut until tree begins to fall. Then stand away*

Tree Removal — 10

BEWARE !
Tree felling can be dangerous!
Keep everyone well away

After felling, cut the tree in pieces and remove them from the roadside.

Now you can remove the tree stump:

- *Dig around the tree*

- *Cut the tree roots around the base of the tree*

- *Tie a rope to the top of the stump and pull*

- *Dig out and remove the roots and stumps*

Remember !
It is sometimes possible to save time by pulling the whole tree over at once.
- *attach a **long** rope to the tree near the top*
- *excavate around the base, and cut side roots*
- *pull the rope, to snap the remaining root below ground level*

You can also use animals to help remove tree stumps.

Removing Small Trees
Remove the whole tree with its roots, rather than cutting it down first and digging out the stump later. This is how to do it:

- *Attach a long rope (check before use) at least 1 m above ground level*

- *Excavate the soil around the roots*

- *Pull the rope to fell the tree - once workers are at a safe distance*

It is better to remove the tree with its all roots; otherwise, deep roots have to be cut by axes. Roots or stumps left in the ground eventually rot leaving holes which weaken the road.

You can also use animal traction to remove small trees with their roots.

11   Boulder Removal

Removing boulders is time consuming and requires an experienced labour force. Consider whether, as an alternative to removing boulders, the level of road can be raised to cover them. But this can be expensive.

If clearance is necessary, consider various methods; either separately or combined:

- *Move the boulders outside the roadway*
- *Bury the boulders on site*
- *Crack the boulders into pieces and remove them*
- *Blast large rocks and remove the pieces*

With boulders up to this size use a crow bar

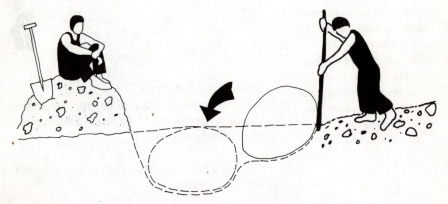

Bury boulders which are more than 0.5 cubic metres or deeply embedded in the ground by digging a hole next to them. First estimate the size of the boulder and consider how large a hole is required. Then dig a hole as close as possible to the boulder and tip it in with crowbars or by using jacks or a winch.

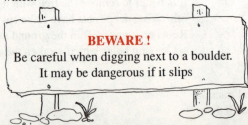

BEWARE !
Be careful when digging next to a boulder.
It may be dangerous if it slips

Boulder Removal

Crack boulders into small pieces for removal if it is not possible to move or bury them. Heat the boulder or rock by starting a fire over it and keeping it going for at least six hours; then cool the rock suddenly by pouring cold water over it; The heat makes the boulder expand; the cold makes it contract and crack so that the boulder can then be broken up with sledge hammers.

This method may be fun to use, but is time consuming and expensive. Only use it as a last resort.

Blasting rocks is an effective method, although dangerous and expensive. Only **specially trained and licensed** supervisors can do this.

Boulders can also be broken up with simple hand tools. Try to find cracks or weak lines. A large rock can be broken up quickly by drilling holes and then splitting the rock with plugs and feathers.

Tools for the job
Use crow bars, shovels, pick axes, sledge hammers, winches, plugs and feathers, wedges, chisels and tongs, as well as safety items such as goggles and gloves.

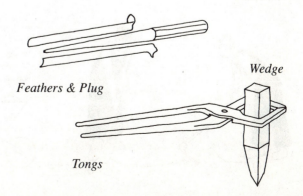

Feathers & Plug

Wedge

Tongs

12 Topsoil Removal

Topsoil is the soil near the surface, in which plants grow. The underlying material is called **subsoil**.

Why remove it?

• *Topsoil contains a lot of organic matter which decays, so it is not a strong layer for building the road on.*

• *Topsoil allows grass and plants to grow through onto the roadway.*

Is removal always necessary?

• *Remove topsoil where it is different from the subsoil.*

• *Leave undisturbed sandy soil, which is a strong layer.*

• *Learn to distinguish between organic and inorganic matter. Organic material includes humus, decayed vegetation and even roots. Organic soil usually smells.*

Tools for the job

Hoes

Shovels

Rakes

Wheelbarrow

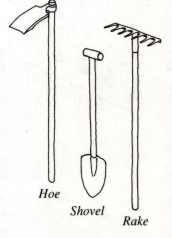

Hoe *Shovel* *Rake*

Topsoil Removal   12

Excavating topsoil

• Dig several holes about 10 m apart, to find out how deep the topsoil is in your working area.

• Calculate the area each person can excavate, using the depth of topsoil measured divided into the volume of the daily task.

• Topsoil is normally loose and easy to excavate, so allow about 5m^3 per person as a daily task.

Pegs mark area to be excavated

What to do with topsoil

• Don't just pile it up on the roadside; it's too valuable.

• Spread it on agricultural land.

• Consider spreading topsoil on the sides of a road embankment, to be grassed later for protection against soil erosion.

13 Formation

The first step in making a good road is **formation**. This is the excavation and shaping necessary to produce the correct cross section and allow room for ditches and side slopes.

How is this to be done ?

First you need to organise the excavation. This is done by setting out the formation, calculating the earthworks and organising the labour force to do the work.

In difficult terrain it may not be possible to set out the levels as explained earlier. The road may have to follow the ground. In this case the road will usually be in steep sidelong ground like this.

Earthworks consist of:

Formation

Slotting

Rock

Earth Moving

Compaction

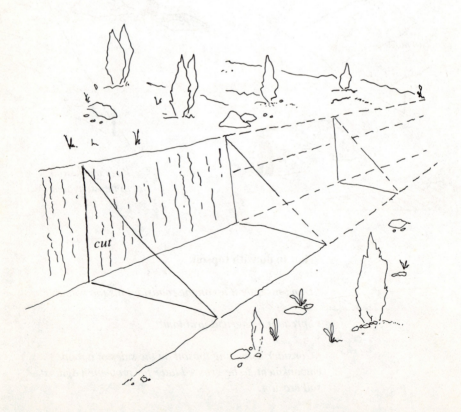

Formation 13

In gently sloping terrain (less than 10% slope) you should choose the alignment to get **equal** cut and fill.

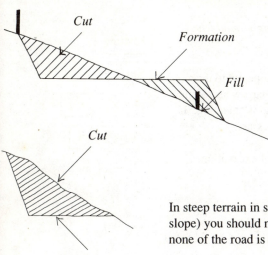

In steep terrain in sidelong ground (more than 10% slope) you should make all the formation in cut, so that none of the road is in fill which would be unstable.

To calculate % slope, divide the change in height of the ground by the distance over which that change has happened. Then multiply the answer by 100.

Example

0.5 m
3 m

Slope =

$$\frac{0.5 \text{ m}}{3.0 \text{ m}} \times 100 = 17\%$$

1 m
14 m

Slope =

$$\frac{1 \text{ m}}{14 \text{ m}} \times 100 = 7\%$$

Remember!
Slope is expressed as a percentage (%)

14   Slotting

The **volume** of work involved in excavating can be calculated by using a method called **slotting**.

• *This is especially important in sidelong ground where excavation will be large.*

Slots are narrow excavations into the slope, up to the setting out peg. They should be cut as narrow as possible every 20 m, or even closer if the ground is changing. In flat ground, of course, slots are not necessary.

You use slots to measure the volume of excavation.

Slot side

20m

After the slot is complete:

• look at the material and see whether it is ordinary soil, or weathered rock, or hard rock

• choose a **task per person** in m^3. For example:

> For hard soil or weathered rock use a task rate of $1.5\ m^3$ per worker day
>
> For a firm soil use a task rate of $3\ m^3$ per worker day
>
> For loose soil use a task rate of $5\ m^3$ per worker day

Slotting   14

- *Measure height and width of slot side and calculate the area of the slot side. Repeat this for the next slot. Now add the two areas and divide by 2 to give you the **average**. Then multiply by the length between the two slots. This is called the **volume of excavation**.*

- *Divide the volume of excavation by the task rate per person to find out how many worker days you need to carry out the work.*

- *Allow a distance of at least 2 m between workers so that they can carry out the task without getting in each other's way.*

To excavate firm soil and weathered rock the method is the same.
Step 1. Excavation should be to a vertical face only.
Step 2. The back slope should be excavated later as a separate task.

Now work out the task on the ground and explain to each labourer what he must do. The excavated material is called **common fill** and must be spread and compacted to the correct cross fall or camber. To do this allow 20 m^3 per worker day.

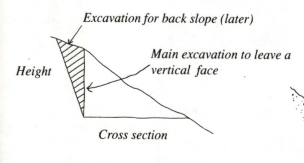

Cross section

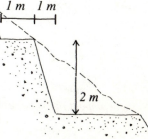

Backslope excavation

Backslopes are usually cut at a slope of 2 vertical: 1 horizontal: so, to calculate the back slope excavation, measure the height of the vertical face and divide by 2 to give you the width. If the height is more than 2 m, make a horizontal step 1 m wide at the 2 m height.

15 Rock

You have already considered excavating ordinary soil and weathered rock. But more difficult and expensive than these two types of ground is **rock** in formation. You can deal with this situation in one of three ways:

• build the road on top of the rock

• excavate the rock with hand tools

• excavate the rock by drilling and blasting

Building on top of the rock

• First set out the road with pegs or paint

• Then, build a masonry wall on each side of the road to retain the fill

• Now bring in common fill between the walls and spread and compact to the correct shape

• If the road is to be gravelled, spread and compact as explained in the gravelling chapter

> The walls are needed to prevent the fill from being washed away.

Rock 15

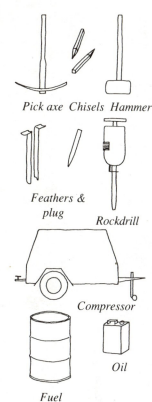

Pick axe Chisels Hammer

Feathers & plug

Rockdrill

Compressor

Oil

Fuel

Excavate the rock with hand tools

• First set the task and explain what each worker should do. Allow 0.5 m^3 of excavation for each worker.

• Always try to keep the rock face vertical as you excavate. Do not try to excavate large pieces of rock, as this will slow down the work and cause problems in moving the material.

• Use feathers and a plug in groups of four to break rock along a line.

• Throw the excavated rock pieces evenly across the road.

Drilling and blasting the rock

Do this only where there are large volumes of rock to be excavated.

• First, make sure the compressor is complete and ready for use, with all tools, oil fuel and so on.

• Then, discuss with the engineer where to drill and to what depth.

• Then make sure that operators are wearing ear protectors, goggles, overalls and boots.

• After drilling the holes, put grass into the top of each hole to keep it clean.

• Arrange for the blasting crew to come and to set out the explosives. Your task is to make sure that the workers are well away from the blasting area.

Operators must wear ear protectors, goggles, overalls, and boots.

16   Earth Moving I

It is sometimes necessary to move common fill along the road to make the road level and smooth. This important task is divided into three parts:

- *loading common fill*

- *hauling or transporting common fill*

- *unloading common fill*

Hauling can be done effectively by light equipment up to a distance of 8 km.

Here is a table that shows the way to haul different materials according to the distance involved:

Hauling distance	Method of haulage	Materials to be moved
0 -10m	Shovelling	Soil
0 - 50m	Stretchers	Stones, soil
10 - 150m	Wheelbarrows	Soil, stone
150- 500m	Animal carts	Soil, stone, water
500- 8000m (8km)	Tractor-trailers	Soil, stone, water

Earth Moving I 16

Loading bay

Loading: Use shovels for loading soil and gravel. Make sure the loading height is convenient - particularly when using trailers. By placing the trailers at the lowest spot, the loading height can be reduced. Prepare loading bays whenever possible.

Hauling: Avoid using a lorry to haul common fill. It is expensive, requires regular maintenance and needs scarce spare parts. Choose the best transport route.

" Ordinary " haul-route for 2 tractors + 4 trailers

Hauling distance (km)	Trips per tractor per day	No. of workers:- Excavate	Load	Spread	Total
0-1	18	36	18	9	63
1-2	14	24	12	6	42
2-3	11	18	9	5	32
3-4	9	14	7	4	25
4-5	7	12	6	3	21
5-6	6	10	5	3	18
6-7	5	8	4	2	14
7-8	4	6	3	2	11

" Poor " haul-route for 2 tractors + 4 trailers

Hauling distance (km)	Trips per tractor per day	No. of Workers:- Excavate	Load	Spread	Total
0-1	18	36	18	9	63
1-2	12	28	14	7	49
2-3	9	22	11	6	39
3-4	7	18	9	5	32
4-5	6	14	7	4	25
5-6	5	12	6	3	21
6-7	4	10	5	3	18
7-8	3	8	4	2	14

17   Earth Moving II

When unloading, loads are dumped close together, beginning at the far end of the fill. When you use tippers or trailers, the fill can be partially spread as the tractor moves forward dumping the load. Some trailers have drop sides which open for unloading.

> **Remember !**
> Dump the loads of fill where they will not block the access for the next loading

Spreading fill

Unloading fill

Working out the tasks

The activities are :
• *Excavating and loading; these are parts of a single activity.*

• *Hauling and unloading; these are parts of a single activity.*

• *Spreading common fill.*

Earth Moving II

The output per worker for hauling and unloading by wheelbarrow will be different for different haul distances.

Example :

Suppose the task for excavating and loading has been set at 5 m^3 per worker day.

Suppose the task for hauling and unloading has been set at 4 m^3 per worker day for a hauling distance (one way) of 30 m, taking into consideration the difficult route.

Suppose the task for spreading has been set at 10 m^3 per worker day.

Take, say, 20 m^3 of common fill to be moved :

20 ÷ 5 = **4** workers are needed for excavating and loading.

20 ÷ 4 = **5** workers are needed for hauling and unloading.

20 ÷ 10 = **2** workers are needed for spreading.

Total = **11**

For every 20 m^3 you need 11 workers.

Remember !
Two wheelbarrows will be needed for each hauler.

Make sure there is a system for measuring each worker's output

Four workers Excavate and load

Five workers Haul and unload

Two workers Spread

18 Compaction

A road which has been **compacted** can stand erosion and traffic better than a loose embankment. Compaction is pressing soil together to make it denser, by getting air out. In this way the soil becomes stronger and more soil particles touch each other.

To achieve the best compaction, you need the right amount of water to "lubricate" the soil particles so that they can compact.

Particles

Voids are filled with air and water

If there is too little water, there is not enough "lubricant" to let the soil particles compact together.

If there is too much water, the soil particles are held apart by the water and the soil becomes soft and cannot be compacted properly.

Voids are filled mostly with water which keeps the particles apart

Compaction can be done by:

• *Natural consolidation, which sometimes can take up to six months, and is not very effective.*

• *Traffic, which may be all that is available for simple roads.*

• *Hand rammers. They must be comfortable to handle, having the right diameter in relation to their weight. A rammer with a diameter of 10cm should weigh 7- 8 kg. The layers of soil being rammed should be less than 15cm thick.*

• *Machines (normally vibrating rollers).*

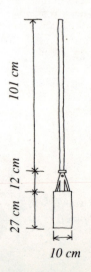

Compaction

Machine Compaction

- Make sure each layer is not more than 15 cm thick, otherwise the roller cannot compact the soil properly.

- As a simple rule the roller should pass over every point of fill a minimun of 8 times and ideally 13 times.

- Only the operator should be allowed to use the roller. If it is not working well, make sure it is reported immediately.

- Compact **along** the line of the road.

- Remember to check the oil and fuel every day.

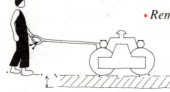

In some places water is not available for watering the soil because getting and transporting it would be too expensive. When this is so, do what you can to prevent the moisture that is in the soil from drying out:

- Spread and compact without delay after transporting.

- Spread and compact in the morning when there may be some humidity.

- Go back and compact at the end of the rainy season.

Only trained operaters should use the roller

Compact in layers 15 cm thick

Roller should make about 10 passes over every point.

19 General

The most important factor in road construction is the provision of a proper drainage system. Excessive water can weaken the road, make it impassable or even wash it away.

Water appears on the ground in different forms:

- **Ground water** *flows through the ground*

- **Surface water** *is found in ditches, streams or rivers*

- **Rain** *(and in some countries, snow or hail) collects and becomes surface water*

So remember that a good road is one which is well drained and allows the water to flow away from it as quickly as possible.

Drainage consists of :

Road Surface Drainage

Side Drainage

Erosion Control

Turnouts

Cross Drainage

Drifts

Culverts

Vented Fords

Bridges

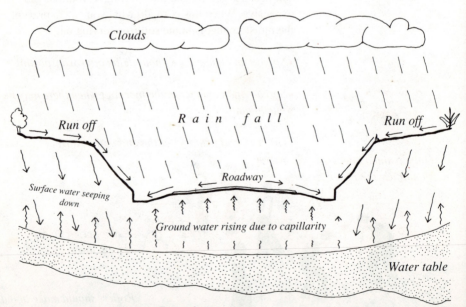

General 19

A good drainage system consists of several components:

• **Surface drainage** *makes the water flow off the road quickly*

• **Side drainage** *carries water from the road, and prevents water from reaching the road*

• **Erosion control** *(or scour checks) slows down the water in the ditches and prevents erosion*

• **A catch water drain** *above a road catches water before it reaches the road*

• **Cross drainage** *allows the water in the side drains to cross the road line by channelling it under or across the road*

• **Turnouts** *take the water in the ditches away from the road (not shown here)*

All these different types of drains have to work together in order to protect the road from being damaged by the flowing water.

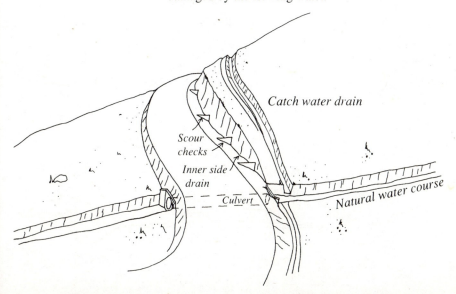

20 Road Surface Drainage

Surface drainage prevents water from damaging the road by leading it off the road quickly. This is done by shaping the road so that the water can flow freely into the side drains.

The slope from either side of the centreline to the sides is called the **camber**. This kind of sloping is used in open terrain, with a ditch on each side. In sidelong ground, with only one ditch, it is better to use a **crossfall**.

Formation is complete

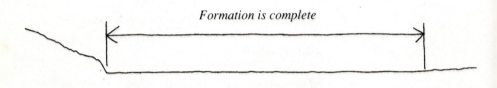

Material from ditches is thrown into centre of road.

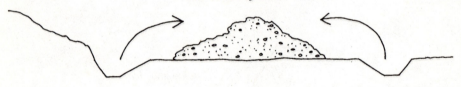

Material is spread to the correct camber (or crossfall)

Road Surface Drainage

For earth and gravel roads, the crossfall or camber should be 5% to 7%. if the slope is 7%, it rises by 7 units for every 100 units of horizontal length.

The camber is made by spreading the soil which has been heaped along the centreline in regular mounds during the formation of the side drains. **Make sure that there is enough soil to make the camber at the correct angle**.

Ensure that the slope is correct by using a camber board and a spirit level. Place the camber board crosswise on the road, with the spirit level on top. Now use the spirit level to get the correct slope.

For a camber board of 7%, the dimensions would be:

Spirit level

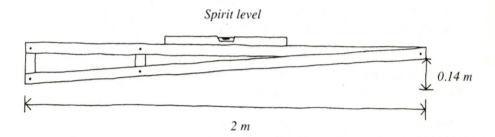

0.14 m

2 m

A line level can be used instead.

Line level

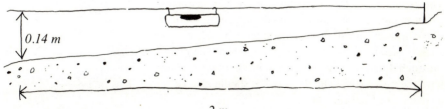

0.14 m

2 m

21 Side Drainage

Side drains or ditches collect water from the carriageway and prevent water from the surrounding areas from reaching the road. The ditches, therefore, have to be large enough to cope with the water.

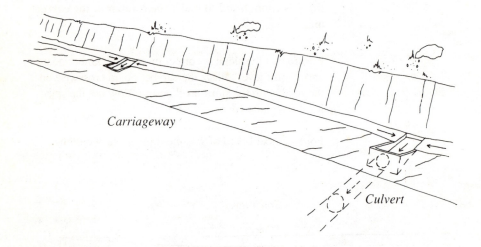

Carriageway

Culvert

The size and shape of the side drain are decided by the engineer. The drain is constructed in two stages:

• **The ditch** is made by digging a rectangular trench. The soil is thrown into the centre of the road.

• **The slopes** are then cut from the edges of the ditch to the base of the trench. Again, the soil is shovelled into the centre of the road.

This two-stage method of making the side drain makes it easier to measure and to work on during construction. The construction will be easier and more accurate if a cut stick is used to measure the correct width and depth of the ditch. Each worker involved in this activity should have their own stick. Ditch templates should also be available for final checking.

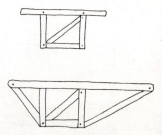

Ditch templates

Side Drainage 21

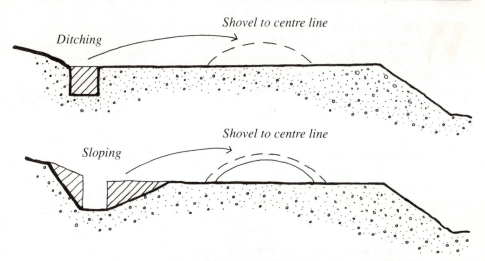

Now you can set the task rate, depending on the type of material to be excavated. You should allow the following amounts per worker day:

Task rates per worker day

Material	Max	Average	Min
Ordinary soil	5.0 m^3	3.5 m^3	3.0 m^3
Weathered rock	2.5 m^3	2.0 m^3	1.5 m^3
Excavated rock	1.0 m^3	0.7 m^3	0.5 m^3

The task includes excavation and throwing the material into the middle of the road.

Now calculate the area of the cross-section of the ditch.

Divide the **task rate** in the table by the **ditch area** to give you the **length** of ditch per worker, per day.

Erosion Control

When the water flows too fast, it can wash away the bottom of the ditch. The faster water flows, the more soil it can erode and carry away. If the ditch becomes too deep, the road may collapse into it. A method of **erosion control** is required to avoid this.

The engineer will advise which is the best method. There are various ways of controlling erosion.

• *Erosion checks, which consist of lines of stones or wooden stakes across the side drain to prevent the soil from washing away.*

• *Grassing along the side drain can strengthen the soil by "knitting" it together with roots.*

Stone erosion checks

• *First cut a furrow about 10 cm deep and about 20 cm wide across the ditch*

• *Then place stones close together in the furrow. Each stone should be about 20 cm across. Do **not** place stones on top of each other as if you were building a wall.*

• *As a general guideline space the checks as follows:*

Steepness	Gradient	Spacing of erosion checks
Very steep hill	More than 10%	5 m spacing
Steep hill	Between 5-10%	10 m spacing
Gentle hill	Between 3-5%	20 m spacing
Almost flat	Up to 3%	Checks are not required

> **Remember !**
> Erosion is caused by fast flowing water

> **Remember !**
> Some soils wash away more quickly then others. Use your own judgement in spacing erosion checks.

Erosion Control

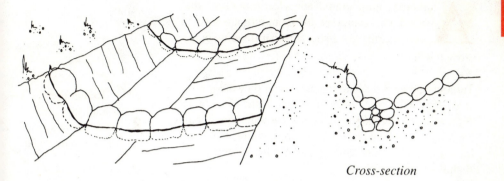

Cross-section

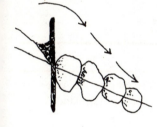

Direction of flow

- Erosion checks can be done as task work but you will have to set a reasonable task rate which will vary from site to site.

Bamboo or wood erosion checks.
These should be driven deeply into the ground. If possible, stones should be placed downstream so that the water falls on a hard surface.

Grassing along the side drain
- First cut a narrow furrow along the ditch and plant grass.

- Water the grass every day for two weeks.

- When setting out the task, allow a 30 m length per worker, per day.

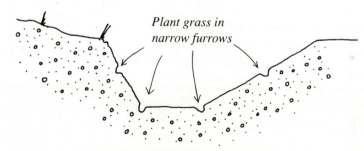

Plant grass in narrow furrows

Furrows

23 Turnouts

A turnout or mitre drain leads water away from the side ditch. Therefore, the more turnouts you provide, the less likely it is that flooding or erosion will occur.

Turnouts are very difficult to get right. If they are too steep, they will cause erosion in the land they lead on to. If they are too flat, they will silt up. Turnouts should have a minimum gradient of 2%. More importantly, they must lead gradually into the land, getting shallower and shallower.

Make sure that the discharged water from the turnouts is channelled towards land boundaries to avoid damage to farm land. If erosion is likely to occur, put a layer of stones on the ground at the end of the mitre drain.

Remember !
Turnouts take water in the ditches away from the roadside

Road gradient %	Mitre drain intervals Should not exceed metres	When discharging the water on to a piece of farmland
12	40	
10	80	
8	120	If exceeded, scouring will occur
6	160	
4	200	
1 - 2	50	If exceeded, silting will occur

(Right column, spanning scouring rows: 20 to 50 metres wherever possible on to a boundary between farms)

Turnouts 23

Drainage

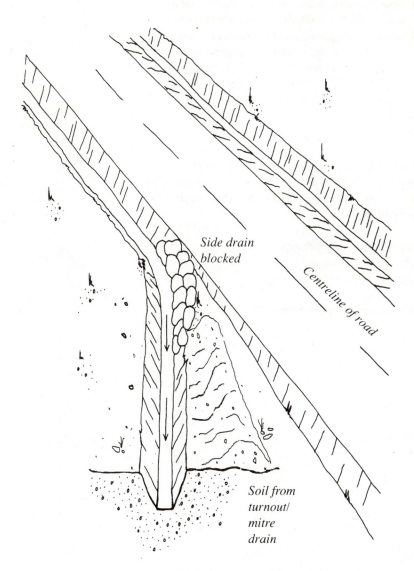

Side drain blocked

Centreline of road

Soil from turnout/ mitre drain

Mitre drain section

Make the angle at which the turnout leads away from the road about 30°

24 Cross Drainage

Cross drainage is almost aways required on a road. This is because it is necessary to lead the water from higher ground on one side of the road to lower ground on the other. Cross drainage is also required where the road goes downhill from both directions to a low point. At this point, water collects and cross drainage is needed to let the water through to the other side of the road.

Remember !
If cross drainage is not provided, then the road may be ruined!

Various structures allow cross drainage to take place. The type used depends on how much water is expected to flow through the structure during a bad flood. How much water is expected depends on :

• *the area of land, or* **catchment**
• *from the amount of rainfall in to the area*
• *the shape of the catchment.*

The most common types of cross drainage

Drift

Drifts allow water to cross the road **on the surface**. Drifts are cheap and easy to construct and are well suited to roads with relatively little traffic.

Cross Drainage   24

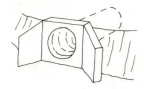

Culverts are usually made of a single line of concrete pipes (normally, not less than 60 cm in diameter to avoid blockage), placed in trenches, and covered over to carry water **under the road**.

Vented fords are usually needed where **large flows of water** are likely. Instead of a single culvert, several pipes are used, surrounded in concrete so that water can flow over the pipes as well as through them without washing them away. **Vented fords** are also called **vented drifts** and **Irish bridges.**

Minor bridges are similar to vented fords, except that pipes are replaced by corrugated iron roofing sheets, surrounded by concrete.

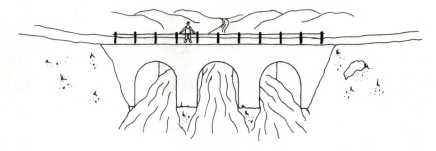

Major bridges are a very complex subject and beyond the scope of this book.

These methods of **cross drainage** are described in more detail on the following pages. All of these involve **concrete making**.

25 Fords (Drifts)

This is how to construct a simple drift:

• Set out the drift according to the dimensions provided by the engineer. The surface of the drift should be at the level of the river bed.

• Excavate for the drift to a depth of 15 cm and place dry stones to a depth of 10 cm.

• Put weldmesh on this bed and place concrete to a total thickness of 15 cm. For mixing and placing concrete, allow a task of 1 m^3 per work day.

• Cover the concrete with sand 25-50 mm thick and then tidy up the site.

• From the following day onwards, water the sand twice a day, for about two weeks.

• Brush off the sand and the drift is ready for traffic.

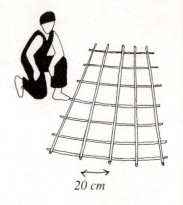

20 cm

Weldmesh is a reinforcing bar welded together in a square grid as shown. For drifts, the bars should be 0.6 cm in diameter and spaced 20 cm apart.

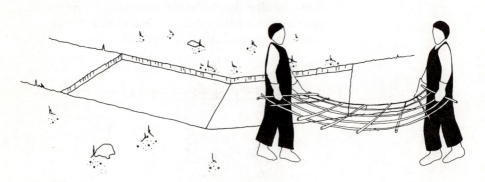

• For small drifts hand packed stone can be used instead of concrete.

Fords (Drifts)

Major drifts can also be constructed and are usually suitable for low-volume roads crossing river beds which are dry for most of the year.

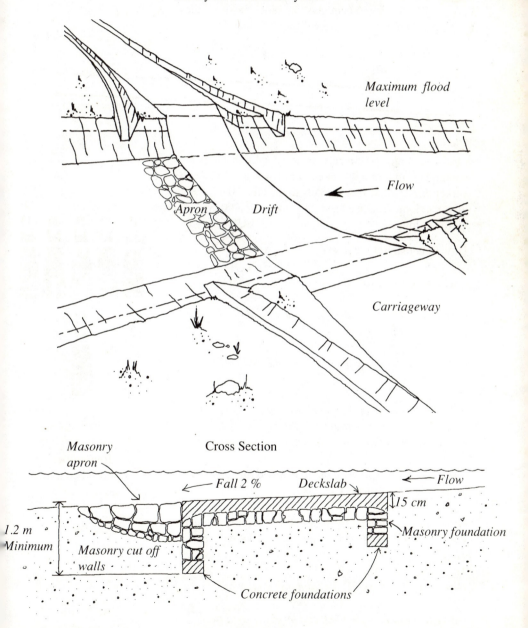

26 Culverts

Concrete pipes can be made on the site, or transported to where they are required. The most common diameter sizes are 60 cm and 90 cm. If they are smaller than 60 cm they can block easily; if they are bigger than 90 cm, they are difficult to handle.

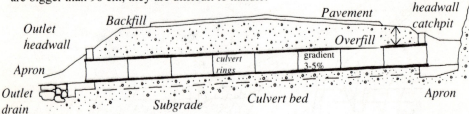

- For installing a 60 cm culvert, allow a 1.6 m run per worker day, and for a 90 cm culvert, allow a 1.3 m run per worker day. It takes five workers to install a 60 cm, culvert up to 8 m long. You need a sixth worker for a 90 cm culvert up to 8 m long.

How to construct a culvert

- Set out and excavate the outlet drain if necessary.

- Excavate a trench. The width should be about $1\frac{1}{2}$ times the diameter of the culvert .

- The depth should be the diameter **plus** the cover shown in the table opposite.

- Form the bottom of the trench so that it is gently sloping (3-5%) in the direction of flow. Now put about 5 cm of loose soil over the whole base of the trench.

- Gently place the pipes in the trench and line them up accurately with crowbars.

- Place the soil around the pipes in layers and compact each layer with rammers. Leave a slight hump over the pipe.

- Now remove the excess soil and tidy up the site.

- After two weeks come back and level the road over the culvert to correct any settlement that has taken place.

Remember !
A gravel bed may be needed in poor or soggy soils

Culverts 26

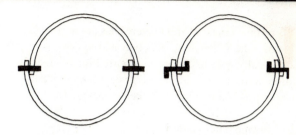

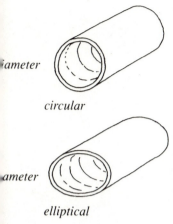

circular

elliptical

Corrugated steel culverts are circular or elliptical pipes which come in separate parts and have to be put together. Pipes come in two semi-circular (or semi-elliptical) parts. Metal links are used to join them together. Because they are lighter than concrete, they can be handled easily and are cheap to transport; **but** they are expensive to buy. You will have to calculate the most economical choice.

Concrete culverts can also be square or rectangular. These can be made on site. They have a larger capacity for the same height as circular ones, but need more concrete to be as strong.

Cover for culverts
It is vital that culverts have enough soil above them to spread the load of traffic.

The following table gives the depth of soil cover and diameter of various types of pipes.

Type of Pipe	Diameter (cm)	Depth of cover (cm)
Spun concrete	60	30
	90	45
	120	60
Concrete cast on site	60	45
	90	70
Corrugated steel (circular) or corrugated steel (elliptical)	60	30
	90	45

27 Vented Fords

In some cases several culvert lines are needed side by side. When more than two are required, it is usually wise to change to a **vented ford**, which is different from a culvert in a very special way : it can be overtopped in a flood without being damaged.

How to construct vented fords

• Set out the position of the vented ford.

• Excavate for the cut off walls and base slab, which is the foundation for the pipes. The base slab should be at the level of the river bed, and should have cross fall of 3% minimum. The cut off walls protect the base slab from being washed away.

• Place the concrete (with weldmesh) for the base slab (10 cm thick).
When the concrete is dry, place the pipes.

• Erect the walls in cement masonry.

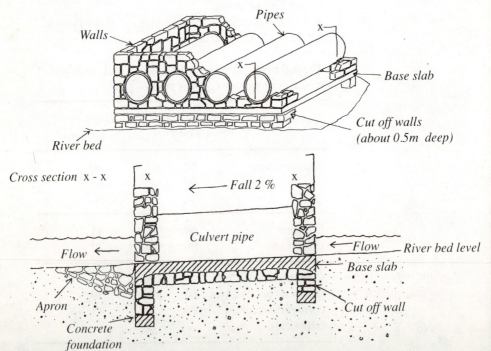

Vented Fords 27

- *Now put lean mix concrete around the pipes.*

- *Erect wing walls of cement masonry as instructed by the engineer. Remember to cut a furrow at least 50 cm deep as a foundation for the wing walls.*

- *Backfill and compact the approaches in layers of 15 cm.*

- *Place the concrete for the slab over the pipes and level off.*

- *Put sand over the concrete and water every day for two weeks.*

- *Construct the apron. The width of the downstream apron should be at least 2 times the height of the vented ford.*

- *After two weeks, the vented ford is ready for use.*

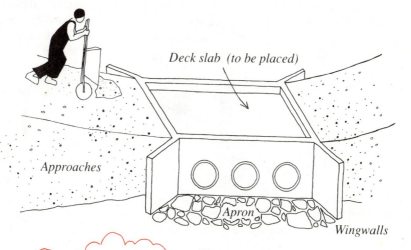

Deck slab (to be placed)
Approaches
Apron
Wingwalls

Remember !
The vented ford deck level **must** be the lowest point in the road so that when there is a flood, the water will flow over the concrete of the vented ford and will **not** wash away the road embankment.

Bridges

Bridges are sometimes necessary to cross major streams and rivers. They are usually costly to construct and may be the weakest parts of the road. The choice of their location is therefore important.

The type of bridge that is used depends on a number of factors:

• *The distance to be crossed*

• *The kind of river to be crossed (seasonal or perennial flow, and maximum and minimum water levels)*

  The amount of traffic expected on the road
• *The cost of construction*

Single and multi-span bridges

The supports on each side of the river bank to take the weight of the bridge are called abutments. If the river is wide, then intermediate supports called piers are required. The distance between the supports is called the span. A bridge without piers is described as being single-span, and a bridge with piers, multi-span.

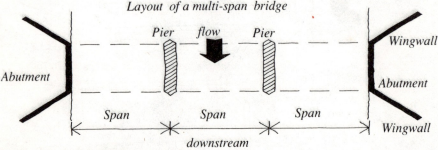

Layout of a multi-span bridge

Various types of bridges are commonly used on rural access roads:

• *Masonry bridges*

• *Timber bridges*

Steel girder bridges and reinforced concrete bridges are usually too expensive for rural access roads.

Bridges 28

An example of a minor masonry bridge

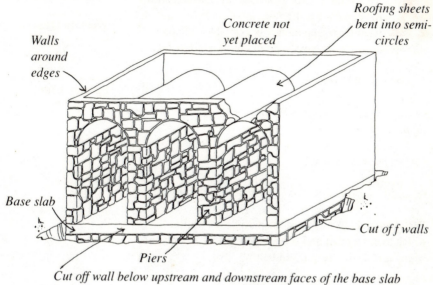

- Base slab 100cm thick on cut off walls
- Piers of masonry or cast in situ concrete with vertical steel
- Arches formed with roofing sheets of galvanised iron bent into semi-circles
- Framework of permanent masonry
- Infill of concrete
- Deck of reinforced concrete

Masonry arches can be built using only masonry but then a timber framework is needed and, **only** skilled masons can build these masonry arches successfully.

An example of a typical timber bridge

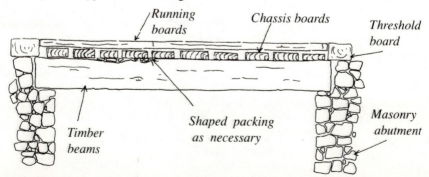

29 General

A road needs a strong surface layer which allows the expected traffic loads to pass in dry and wet weather. When money is available, the most appropriate surface layer is one of **compacted gravel**.

A good gravel is one which:

• *Remains strong and does not become slippery, even when wet*

• *Does not wash away*

• *Can be excavated by hand (for labour-based projects)*

The strength comes from stone particles of various sizes which lock together and spread the traffic load on to a larger area of the natural soil below.

A good gravel should have a mixture of stones, sand and clay in roughly the following proportions :

Gravelling consists of :

| Work in Quarries |
| Haul, Unload |
| Spread, Compact |

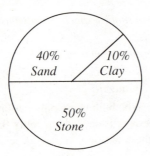

Gravel contains:
• **Stones** for strength
• **Sand** to fill the gaps between the stones
• **Clay** to bind the stones and sand

Gravelled roads are stronger and last longer than earth roads. However, gravelling is expensive, because it requires excavation, transport, spreading and compaction.

Because gravelling is expensive, it is very important to find good gravel. Ideally samples of gravel should be tested in a laboratory to make sure it has the right properties.

Remember !
In dry areas, a higher proportion of clay is needed to bind the stones

General 29

- *A place where suitable gravel is found is called a **quarry**. Preferably, this should be as near the road as possible. Finding gravel is a very skilled business and should be done before work starts on the road.*

- *The gravel is **excavated and loaded** by hand, usually onto tippers or tractor/trailers. Wheelbarrows can be used if the distance is less than about 100 m. In some countries animal haulage is suitable.*

- *The tippers or tractor/trailers transport the gravel to where it is needed.*

- *The gravel is **spread** by hand to the right thickness and is then **watered** for easy compaction.*

Towed bowser

Gravel is wetted

- *While still wet, the gravel is **compacted**, ideally using a vibrating roller.*

Work in Quarries

A good quarry is one which requires the **minimum work** for the **maximum output**. In selecting a quarry a number of aspects should be considered:

- *The quality of the material*

- *The depth of soil over the gravel material*

- *The cost of excavating and transporting the material*

- *The land owner of the quarry*

- *The hauling distance*

- *The need for an access road*

> **Remember !**
> To start up a quarry:
> - decide on layout
> - remove topsoil
> - excavate loading bays

The quarry layout should allow the vehicles and carts to enter and leave without being in each other's way. A circular traffic flow, requiring only single lanes is ideal. If only a single access road is possible, then allow a double lane for traffic in both directions.

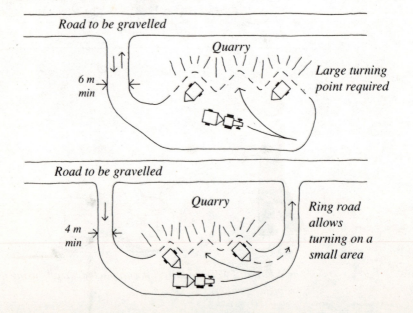

Work in Quarries 30

Gravelling

In the Quarry

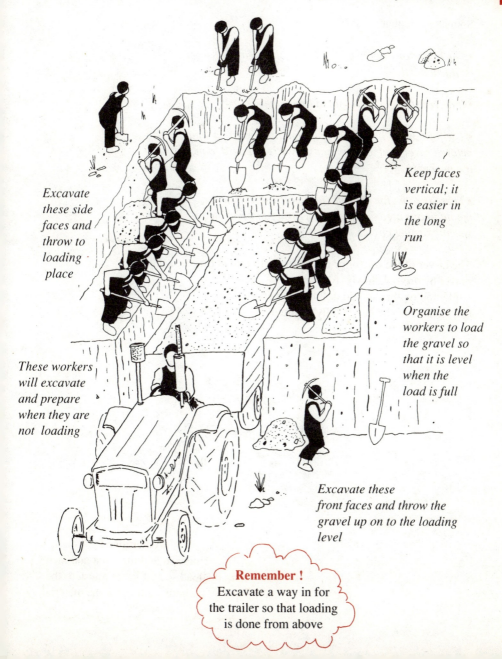

Excavate these back and side faces above with crow bars .... and below with pick axes

Keep faces vertical; it is easier in the long run

Excavate these side faces and throw to loading place

Organise the workers to load the gravel so that it is level when the load is full

These workers will excavate and prepare when they are not loading

Excavate these front faces and throw the gravel up on to the loading level

Remember !
Excavate a way in for the trailer so that loading is done from above

31   Haul, Unload, Spread, Compact

Hauling is expensive. The tippers or tractor/trailers are a very expensive part of the operation.

So make sure that:

• *The tipper or tractor/trailer leaves the quarry as soon as it is loaded.*

• *It drives straight to the spreading gang.*

• *The formation is already prepared to the correct shape before gravel is placed so that gravel is **not** used to make a camber.*

• *There is room for the vehicles to turn safely. In mountainous areas it may be necessary to excavate special turning circles.*

The driver of the tractor or tipper can help spread the gravel by driving **slowly** forward while tipping, but should not turn corners or drive fast until the bed is fully lowered.

Remember!
The vehicle should turn first and then tip the load so that it can return to the quarry without waiting for the material to be spread.

Haul, Unload, Spread, Compact

Before the next load is tipped, make sure that the formation has been prepared to the correct crossfall or camber.

The number of workers needed to spread and shape the gravel depends on the volume of gravel delivered during the day.

Generally two workers are needed to collect and spread water.

One worker is needed to break up the large stones with a hammer.

Raking is important. Good raking gives a good surface.

The roller must work all the time : as a guide, for gravel in layers of 15 cm, you need 8 to 13 passes over each point.

Principles

In order to achieve objectives, you have to plan activities in advance. Before detailed plans can be made, the sequence of activities must be understood. Here is an example of construction activities carried out in sequence:

Planning consists of:

Principles

Daily Work Plan

- *Supporting*
  - Setting out work at camp
  - Water supply to the camp

- *Site Clearing*
  - Bush clearing
  - Topsoil / root removal
  - Tree and stump removal
  - Boulder removal

- *Earthworks*
  - Excavation and filling
  - Spreading and compaction

- *Drainage*
  - Ditch excavation
  - Culverts and structures
  - Erosion protection

- *Gravelling*
  - Excavating and loading
  - Hauling and unloading
  - Spreading, watering and compacting

Remember !
Always plan each activity to follow the previous one as closely as possible. Workers from different gangs should not be in each other's way.

Principles

In the same way, administration activities must also be carefully planned and carried out in sequence. Particular attention should be given to planning the recruitment of workers. The engineer will first estimate the overall numbers of labourers needed for the work.

By following the construction activities in sequence, a plan for taking on workers can then be made. For example:

- *Week 1 : Gang A starts with 25 workers on supporting activities and clearing*

- *Week 2 : Gang B starts with 50 workers on the earthworks*

- *Week 3 : Gang C starts with 25 workers on the drainage*

- *Week 4 : Gang D starts with 75 workers on gravelling*

When the sequence of events has been understood, it is time to think about the details of each activity.

The answers to these questions should be found in **work plans** which are completed on a daily, weekly and monthly basis.

33 Daily Work Plan I

The **daily work plan** is the most detailed of all the plans. It is an essential tool for supervisors, since it shows specific targets to be met in an actual day:

- *Which activities are to be carried out*

- *Where they are to be carried out*

- *How many workers are to work on each activity*

- *How much work they should do*

The supervisor prepares and carries out the daily work plan on the following basis:

- *The measured **quantities of work** (such as topsoil removal, excavation, ditching, sloping and camber formation)*

- *The estimated **task rate***

- *The **number of workers** available*

- **First** the supervisor should **measure the amount of work required.** This should be done well ahead of construction especially for excavation, which is a time-consuming activity to measure.

- **Then, the amount of work each worker is expected to do in a day** has to be estimated. The engineer issues task rates for different conditions. The supervisor should observe how work is progressing and judge which daily task is reasonable.

- **Finally, the number of workers** should be determined according to their availability and the extent of the site. The supervisor should be able to work out how many are available for the various activities.

Daily Work Plan I   33

Planning

Part of a daily work plan for an earthworks gang may look like this:-

DAILY* / ~~WEEKLY*~~ / ~~MONTHLY*~~	WORK PLAN* / ~~PRODUCTION RECORD*~~			*Delete as necessary		
SITE:	Date of work **11/8/89**	OFFICER:			Date of form **10/8/89**	

ACTIVITY	Standard Task Rate	Gang Reference	Output	Task Rate	Input	Comments
Excavate ORDINARY SOIL	5.0 m³/wd	B	153	4.5	34	very dry
Excavate HARD SOIL/GRAVEL	2.5 m³/wd					
Excavate ROCK	0.5 m³/wd	B	4	0.5	8	
Excavate & Load COMMON FILL	5.0 m³/wd					
Haul Common Fill Up To 200 metres	5.0 m³/wd					
Spread Common Fill	20 m³/wd	B	157	20	8	
Compact GRAVEL-FILL	250 m³/rd	B	157	157	1	

LOCATION OF ACTIVITIES   Direction of increasing chainage ———▶

Clearing
Formation
Ditches
Gravelling

CHAINAGE 1.5   1.6   1.7   1.8   1.9   2.0   2.1   2.2   2.3   2.4

Totals: ...**50**... wds ...**1**... rds .... tds   Approved by: ....   Date: ....

☁ **Remember !**
Take your completed work plan with you to site

- The supervisor calculates that there are 4 m³ of hard rock to excavate.

- A standard task rate of 0.5 m³/wd is chosen to excavate the rock, but for the ordinary soil the rate is reduced from 5.0 m³/wd to 4.5 m³/wd because the soil is a little hard.

- Up to 50 workers are available.

8 workers will be needed to excavate the rock.
If **34 workers** excavate ordinary soil then a total of 157 m³ of soil and rock will have to be spread. This will need 157/20 = **8 more workers, giving a total of 50 workers.**
For compaction one extra worker is required to operate the roller.

34 Daily Work Plan II

For gravelling activities, machines as well as people must be kept busy.

Example : **Gravelling plans :**

Number of tippers available for this quarry	2
Capacity of tippers	6 m^3
Number of loading bays open	2
Number of workers available	60 to 70
Expected journey time	15 minutes
Typical task rate for quarry gang	2.5 m^3/wd
Typical task rate for spreading gang	10 m^3/wd

> **Remember !**
> The size of gang will also vary with site conditions.

Rough calculations :

- For short haul distance, 1 tipper per loading bay seems right
- Table below shows that each tipper could do 12 trips per day
- Volume of gravel required = 2 x 12 x 6 = 144 m^3
- Table below shows numbers required at quarry = 2 x 24 = 48 workers
- Numbers required to spread and compact gravel = 144/10 = 14 workers

Total of 48 + 14 = 62 workers will be required.

Journey times for 6m^3 tipper

		\multicolumn{7}{c}{Group task: Number of loads}
Journey time		up to 15 mins	15-20 mins	20-25 mins	25-30 mins	30-35 mins	35-40 mins	40-45 mins
Number of workers	29	15	*14*	12	10	*	*	*
	28	15	*14*	12	10	9	*	*
	27	14	*13*	11	9	8	8	*
	26	13	*12*	11	9	8	7	6
	24	13	*12*	10	9	8	7	6
	23	12	11	10	9	8	7	6
	22	11	11	9	8	7	6	5
	21	11	10	9	8	7	6	5
	20	10	9	8	7	6	5	5

Daily Work Plan II 34

Planning

Part of a daily work plan for a gravelling gang may look like this:-

DAILY* / WEEKLY* / MONTHLY*	WORK PLAN* / PRODUCTION RECORD*		*Delete as necessary			
SITE:	Date of work **11/8/89**	OFFICER:			Date of form **10/8/89**	
ACTIVITY	Standard Task Rate	Gang Reference	Output	Task Rate	Input	Comments
Excavate TOPSOIL	5.0 m^3/wd					
Excavate HARD SOIL/GRAVEL	2.5 m^3/wd	D 1	144 *	2.8	48	* Loose Volume
Spread Water and Compact GRAVEL	10 m^3/wd	D 2	144	10	14	Task in-cludes breaking lumps
Haul GRAVEL 0 - 2 km	m^3/wd					
Haul GRAVEL 2 - 7 km	m^3/wd	D	144	72	2	Quarry 0.5 km from km 0.0
Haul GRAVEL 7 - 20 km	m^3/wd					
Compact GRAVEL/FILL	250 m^3/rd	D	144	144	1	

LOCATION OF ACTIVITIES — Direction of increasing chainage →

Clearing
Formation
Ditches
Gravelling

CHAINAGE: **1.5** **1.6** **1.7** **1.8** **1.9** **2.0** **2.1** **2.2** **2.3** **2.4**

Totals: **62** wds **1** rds **2** tds Approved by: .... Date: ......

This plan shows that the gravelling is catching up with the formation gang. It may soon be necessary to :-

• Start a second earthworks gang

• Transfer some people from gravelling gang D to formation gang B

• Use one tipper to haul materials for a few days

• Off - hire one tipper

These decisions should be referred in good time to the engineer.

Remember !
A plan that has to be changed is better then no plan at all.

35 General

At all times you must be in control of the work, so that you can make sure you complete the work on time, and correctly, to the appropriate required standard.

Reporting and Control consists of :

Production Records

Examples of Production Records

There are various types of control:

- *Production*
- *Quality*
- *Cost*

At site level, **production and quality controls** are the most important.

Production controls involve:
- **Inputs**, *or what is needed to do a job (number of workers, number of days, amount of material)*

Number of workers   *Number of days*   *Amount of material*

- **Outputs,** *or what is produced: for example, lengths, area or volume*

Road length

Area cleared

Volume excavated

General   35

A **task** is the amount of work given to a person for one day's pay. The incentive is that the worker can go when he or she has finished the task set.

Piecework is organised differently; workers are paid according to how much they produce. The incentive is that the harder a worker works, the more money he or she earns. Piecework is more complicated to manage than taskwork.

Daywork is payment for attendance not related to output. There is no incentive for improving productivity in daywork, as the worker cannot finish early (as in taskwork) or earn more (as in piecework).

Worker   Day

The amount of work a person does in one day

Control is kept by means of regular inspection. Work results are recorded in reports which are classified in various ways:
- **By time** *(eg daily, weekly, or monthly.)*

- **By activity** *(eg excavating, hauling, or erecting cement masonry.)*

- **By item** *(eg vehicles, tools, or stores.)*

If you are a **supervisor**, you have to deal with all these types of reports. An accurate report can only be made by someone who knows how to organise the site and who regularly inspects it.
Reporting gives essential information to those who planned the work, and who are controlling the finance and expenditure.

An important reason for reporting is that it helps to improve planning and costing next time.

> **Remember !**
> Jobs in construction which are simple and repeatable are called **Activities**.
> Everything done on site can be subdivided into **Activities**.

36 Production Records I

The way to obtain information from the site is first by inspecting the daily work. Before the workers leave at the end of the day, the work carried out should be inspected and approved.

The duties of a supervisor are to:

• *Set out the work to be done before the labourers arrive*

• *Explain the work to be done to the labourers and gang leaders*

• *Supervise the labourers at work and help with any problems*

• *Check whether the right amount of work has been finished*

• *Check whether the work is up to the correct standard*

• *Check that the correct number of workers have been engaged in the work*

• *Set out the next section of work*

Remember !
Explain the work carefully to the workers. Good management means:
**being kind
being firm
being fair**

Production Records I

If anything is not up to standard, find out the reason and give instructions to improve the situation.

Once the work is approved then the workers can leave. But the leader must be told what needs to be done the following day.

Now in the daily record book write the **input** for each activity.

The daily record is a sort of diary in which a number of items are noted every day:

• **The inputs** *to each activity in worker days*

• **The location** *of the different activities carried out*

• **The number** *of workers carrying out the various activities*

• **The plant,** *or vehicles, used on the site*

• **Accidents** *and unusual happenings*

The daily records should be summarised each week. This will simplify the writing of the monthly report.

37 Production Records II

A **daily** production record should allow information about the activities on site to be recorded in a systematic way every day. The form shown opposite can be used to do this and can also be used to summarise information over longer periods.

This **daily production record** looks exactly like the **daily work plan**, but records what **actually happened** rather then what was **planned to happen**.

The main activities are listed, together with **standard task rates**.

The **units** given for these task rates are:

$$\frac{\text{Units of output}}{\text{Units of input}}$$

For example, the **standard task rate** for excavating topsoil is :

$$5.0 \text{ m}^3/\text{wd}$$

Which shows that the **output** is measured in m^3, and the **input** is measured in **wds.**

Units of output :-

m^3	cubic metre
m^2	square metre
lm	linear metre
No	number

Units of input :-

wd	workerday
td	tipperday or tractorday
rd	rollerday

During the day the supervisor writes down the **actual input** for each activity. At the end of the day the **actual output** for each activity is then measured. The supervisor then calculates the **actual task rate** or **productivity** by **dividing the actual output by the actual input.**

For example, if a gang has 10 workers, and they excavate 23 m^3 of soil, the **productivity** is

$$\frac{23 \text{ m}^3}{10 \text{ wds}} = 2.3 \text{ m}^3/\text{wd}$$

A completed form is of little value unless it clearly shows the **date** and **gang** it refers to, and the **location** of the work that has been done.

Production Records II

A typical Production Record looks like this:

Daily*/ Weekly*/ Monthly*	*Work Plan / *Production Record			*Delete as necessary		
Site:...............	Date of work......	Officer:................			Date of form......	
ACTIVITY	Standard Task Rate	Gang Reference	Output	Task Rate	Input	Comments
Clear BUSH	m^2 / wd					
Excavate TOPSOIL	5.0 m^3/wd					
Excavate ORDINARY SOIL	5.0 m^3/wd					
Excavate HARD SOIL/GRAVEL	2.5 m^3/wd					
Excavate ROCK	0.5 m^3/wd					
Excavate & Load Common Fill	5.0 m^3/wd					
Haul Common Fill	5.0 m^3/ wd					
Spread GRAVEL or Common Fill	m^3/ wd					
Install 600 mm CULVERT	1.5 linm / wd					
Install 900 mm CULVERT	1.0 linm / wd					
Erosion checks	10 No / wd					
Mix and Place Concrete	1.0 m^3 / wd					
Erect Concrete Masonry	1.0 m^3 / wd					
Haul GRAVEL 0 - 2 km	m^3/ td					
Haul GRAVEL 2 - 7 km	m^3/ td					
Compact GRAVEL / FILL	m^3/ rd					
Dayworks	wd					

LOCATION OF ACTIVITIES   Direction of increasing chainage →

- Clearing
- Formation
- Ditches
- Gravelling

CHAINAGE

Totals: .... wds   .... rds   ...... tds   Approved by: ....   Date: ......

Reporting & Control

38 Examples Of Production Records

Earthworks Example

Suppose you are in charge of the 26 workers in gang **A**. Your senior officer has told you to strip topsoil and has shown you where. The formation width is 9.2 m.

What do you do ???

• You ought to know how deep the topsoil is. If you have not already found out, quickly dig a few holes to find out the depth of the topsoil (see chapter on earthworks).

• Suppose the depth is 0.2 m, on average.
The width is 9.2 m
The average depth is 0.2 m
The **task** per worker is **5.0 m^3 per wd**
So the **length** per worker is

$$\frac{5.0 \text{ m}^3}{9.2 \text{ m} \times 0.2 \text{ m}} = 2.7 \text{ m}$$

• Mark out lengths of 2.7 m along the road and **show each worker their task.**

• As each worker finishes, check they have excavated all their topsoil **and** removed it from the formation width.

• Stay at the site until all the workers have finished their work and all the tools have been checked and collected.

• Fill in the **Daily Production Record.**

The **input** is 26 workerdays if every task has been completed, the **output** is $26 \times 9.2 \text{ m} \times 0.2 \text{ m}^3 \times 2.72 \text{ m} = 130 \text{ m}^3$

The form is filled in like this :-

ACTIVITY	Standard Task Rate	Gang Reference	Output	Task Rate	Input	Comments
Excavate TOPSOIL	5.0 m^3/wd	A	130	5.0	26	

Examples Of Production Records — 38

Gravelling Example

Suppose you are in charge of gang **D2**, which **spreads, waters** and **compacts gravel**. According to your daily work plan you are expecting 2 tippers to deliver a total of 144 m^3 of gravel.
At the start of the day you have all that you need, but only 13 workers arrive, one fewer then expected.
What do you do ???

- Before the first load of gravel arrives, you must tell each worker what to do. Most workers will spread gravel, but some will have to do other special activities. You may choose to divide up your gang as follows

> **Remember !**
> When things go wrong you have a **challenge**, not a **problem.**

ACTIVITY	NUMBERS
Collect & spread water	3
Break stones	1
Operate roller	1
Spread gravel	8

- Use pegs and string to show where the edge of the gravel should be, and mark where each load should be tipped**.**

- During the course of the day, one tipper breaks down after delivering only 6 loads. The other tipper delivers 12 loads as planned.

- When all the work has been checked, tools collected, fill in the **Daily Production Record.**

The **inputs** are : 13 workers one of whom operates the roller and 1 rollerday

The **outputs** are : 18 loads x 6 m^3 per load = 108 m^3 gravel

The actual task rate is $\dfrac{108 \text{ m}^3}{13 \text{ wd}}$ = 8.3 m^3/ wd

The form is filled in like this :-

ACTIVITY	Standard Task Rate	Gang Reference	Output	Task Rate	Input	Comments
Spread water and compact GRAVEL	10 m^3 / wd	D 2	108	8.3	13	Tipper C1680 broke down at 11.30 am
Compact GRAVEL	250 m^3/ rd	D 2	108	108	1	

39 Tools

The requirements for **tools** depend on various factors:

- The size of the labour force

- The types of soil

- The type of terrain

- What work is required

Stores consists of:

Tools

Materials

In a situation where 70-80 workers are working on a small road project the following tools may be required:

Tools	Number	Spare Handles
Hoes	50	5
Mattocks	10	2
Forked hoes	25	5
Pick axes	15	2
Shovels	50	5
Crowbars	5	
Wheelbarrows (preferably with pneumatic tyres)	10	
Bush knives	30	
Slashers	10	
Axes	10	5
Earth-rammers	10	2
Rakes	10	2
Buckets	3	

There are also tools required in stony or rocky terrain:

- *Sledge hammers*
- *Wedges*
- *Plug*
- *Feathers*
- *Chisel and tongs*
- *Safety glasses*

Remember !
Good quality tools make it easier to do good quality work.

Tools

Miscellaneous items:

2	Oilstones
1	Camber board
1	Straight edge
1	Spirit level
2	30 m tape measures
1	Abney level
10	Boning rods
3	Line levels
10	Ranging rods
1	First-aid kit
	Equipment for watering

Tools should be issued every morning and returned in the evening. The daily control of tools is the responsibility of the storekeeper. If there is no storekeeper, the responsibility lies with the supervisor.

An inventory of tools involves checking the number of tools and their state of repair. Each week the tools should be counted and the actual number compared to those in the store ledger.

Make sure your tools are sharp and in good condition, as well as making sure they are all there.

Take care in storing tools and equipment:

• *Stack items neatly, so that they can be counted easily*

• *Stack different items separately*

• *Stack items of different sizes separately*

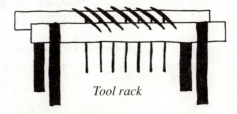

Tool rack

Materials

The supply of materials involves various tasks:

• *The keeping of a store*

• *The maintenance of appropriate records*

• *The ordering of certain types of materials from headquarters*

• *The local purchase of materials such as sand, stone (aggregate), or timber*

Stores are divided into **Permanent** and **Consumables**

The construction materials needed include the following consumables :

• Culvert rings

• Tarpaper *(for covering the culvert joints)*

• Stone

• Sand

• Cement

• Gabion baskets

• Marker posts *(for culverts and structures)*

Remember !
- **Permanent** stores have to be returned
- **Consumables** are used up
- **All** stores issues must be recorded

Some of these items might not be brought to the store, but directly to the place where they are required.

The supervisor must always ensure that all stores are inspected and entered in the **stores - on- site** book.

Materials   40

In addition to the tools (which are permanent stores) and construction materials (which are consumables) described opposite, other items, including the following, may be needed:

• **String, posts** and **paint** f*or setting out (consumables)*

• **Ropes** *for culvert laying (permanent)*

• **Nails and sawn timber** *for repair of huts (consumables)*

• **Wood preservatives** *for impregnation of wood in huts or bridges (consumables)*

• **Torch** *for the night watchman (permanent)*

• **Batteries** *for the torch (consumables)*

• **Coat and blanket** *also for the watchman (permanent)*

• **Water containers** *for drinking water (permanent)*

• **Stationery**, *such as muster rolls, blank paper, pens, pencils, and duplicate books (consumables)*

Remember !
Never sign for consumables as permanent stores

This is how to store materials:

• *Stack fuel and paraffin inside, away from other stores*
• *Stack cement on a well ventilated timber floor, in a separate hut*
• *Stack other items separately and neatly*
• *Make sure you have a watchman on duty at all times*

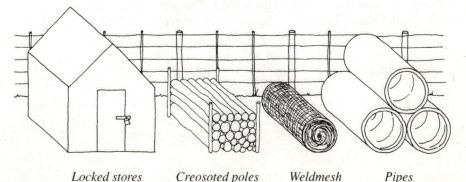

Locked stores   Creosoted poles   Weldmesh   Pipes

41 Setting up Site

The site is under the responsibility of the **supervisor**, who receives instructions from the **engineer**. The supervisor is responsible for:

- The camp
- The stores
- The resources

Starting and Organising consists of :

Setting up Site

Camp Layout

The site must be well organised and arranged from the very beginning.

There are four important elements that make a site work well:

- The location and layout of the **camp**

- The availability of a sufficient quantity of **stores**

- The availability and organisation of **labourers** at the right time

- The good **administration** of site matters

Always remember that **labourers** cannot be hired, until the **camp, tools, equipment and plant** are on site and until the **staff** are available.

To help you remember the right sequence of events, follow the route shown opposite.

Control of stores must start from the first day. This means that a **watchman** must be on site, and that a **storekeeper** should be responsible for recording all deliveries.

The **time required** to set up a new site can vary from:
- **a few days**, if everything is available, to
- **many months**, if orders have to be placed and staff recruited and trained.

Setting up Site 41

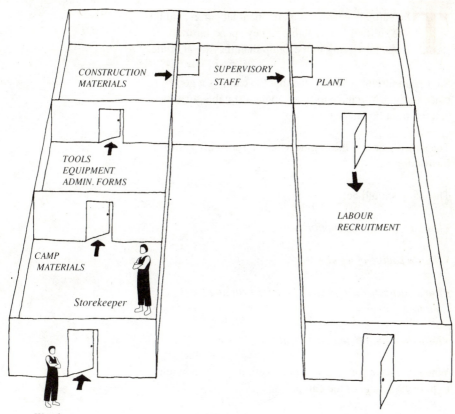

Before the job can begin, you need to complete the items shown in sequence. Labour recruitment can only start when all the other items are complete.

Remember !
Everyone should know that the recruitment is fair. Make sure that :
- All adults who may wish to work are told about the recruitment well in advance.
- A public ballot is used to select the required number of workers.

Labour recruitment should be done gradually: don't employ everyone at once but rather over several weeks so that everyone can learn the job.

Starting & Organising

42   Camp Layout

The camp is where any permanent labourers and the site staff all live together. They spend far more of their time living in the camp than they do at home.

So it is important for everyone that the camp is a good place to live. You have to think about a suitable **location** and the right layout of the camp.

The supervisor and the engineer select a **suitable camp location**.

The camp should be:

• On the actual construction site

• Near a source of good water

• Easily accessible to project vehicles bringing the various stores

• Near, but not too near, the local community

When you have found a suitable location, you can begin to plan the layout of the camp.

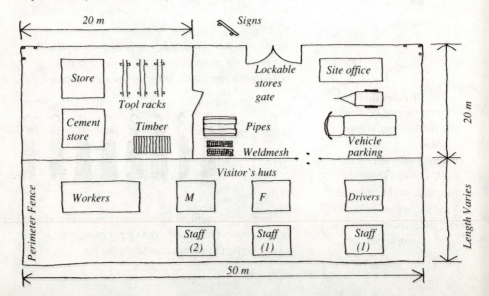

Camp Layout

Each camp is different, and some may be considerably smaller than the example shown. Every camp will, however, need the following:

• Accommodation and equipment for staff

• Hooks, padlocks and chains

• Office furniture and stationery

• Timber for stores shelving

• A supply of clean water (allow at least eight litres per person per day)

• A pit latrine (should be at least 3 metres deep and at least 25 metres away from the camp)

Larger camps may also need:

• Poles, standards, barbed wire and staples for camp fencing

• Radio, aerial and batteries

• Signs and poles

Pit latrine

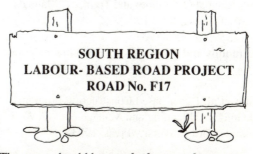

The camp should be **regularly moved** so as to remain close to the work on site. As a rough guide the distance between successive camp sites should be between 5 and 10 km.

References

• **Transportation Research Board, National Research Council: -** Washington DC, USA:
- Synthesis 1: Maintenance of unpaved roads (1979)
- Synthesis 2: Stage construction (1980)
- Synthesis 3: Labour-based Construction and Maintenance of Low-Volume Roads (1981)
(Washington, TRB).

• **G.A. Edmonds and J.D.G.F. Howe (eds):** Roads and Resources: Appropriate technology in road construction in developing countries (London, Intermediate Technology Publications, 1980).

• **J.J. de Veen:** The rural access roads programme: Appropriate technology in Kenya (Geneva, ILO, 1980).

• Guide to tools and equipment for labour-based road construction (Geneva, ILO, 1981).

• **G.A. Edmonds and J.J. de Veen:** The application of appropriate technology in road construction and maintenance: A learning methodology (Geneva, ILO, 1981).

• **L.S. Karlsson and J.J. de Veen:** Guide to the training of supervisors for labour-based road construction and maintenance Instructor's Manual (1 volume) and Trainees' Manuals (2 volumes). (Geneva, ILO, 1981).

• **G.A. Edmonds and J.J. de Veen:** Road maintenance: Options for improvement (Geneva, ILO, 1982).

• **B. Coukis et al:** Labor-based Construction Programs: A Practical Guide for Planning and Management (A World Bank Publication: Oxford University Press, 1983).

• **J. Hindson: revised by J. Howe and G. Hathway:** Earth roads - Their construction and maintenance (London, Intermediate Technology Publications, 1983).

References

- **K. G. Vaidya:** Guide to the assessment of rural labour supply for labour-based construction projects (Geneva, ILO, 1983).

- **United Nations Economic Commission for Africa, Addis Ababa, Ethiopia:** - Road Maintenance Handbook:
Volume 1: Maintenance of roadside areas, drainage structures and traffic control devices
Volume 2: Maintenance of unpaved roads
Volume 3: Maintenance of paved roads
(Addis Ababa, ECA, 1982).

- **S. Hagen and C. Relf :** The district road improvement and maintenance programme - Better roads and job creation in Malawi (Geneva, ILO, 1988).

- International course for engineers and managers of labour-based road construction and maintenance programmes. Course notes (2 volumes); Course curricula and lecturer's notes (1 volume). (Geneva, Infrastructure and Rural Works Branch, ILO, in preparation).